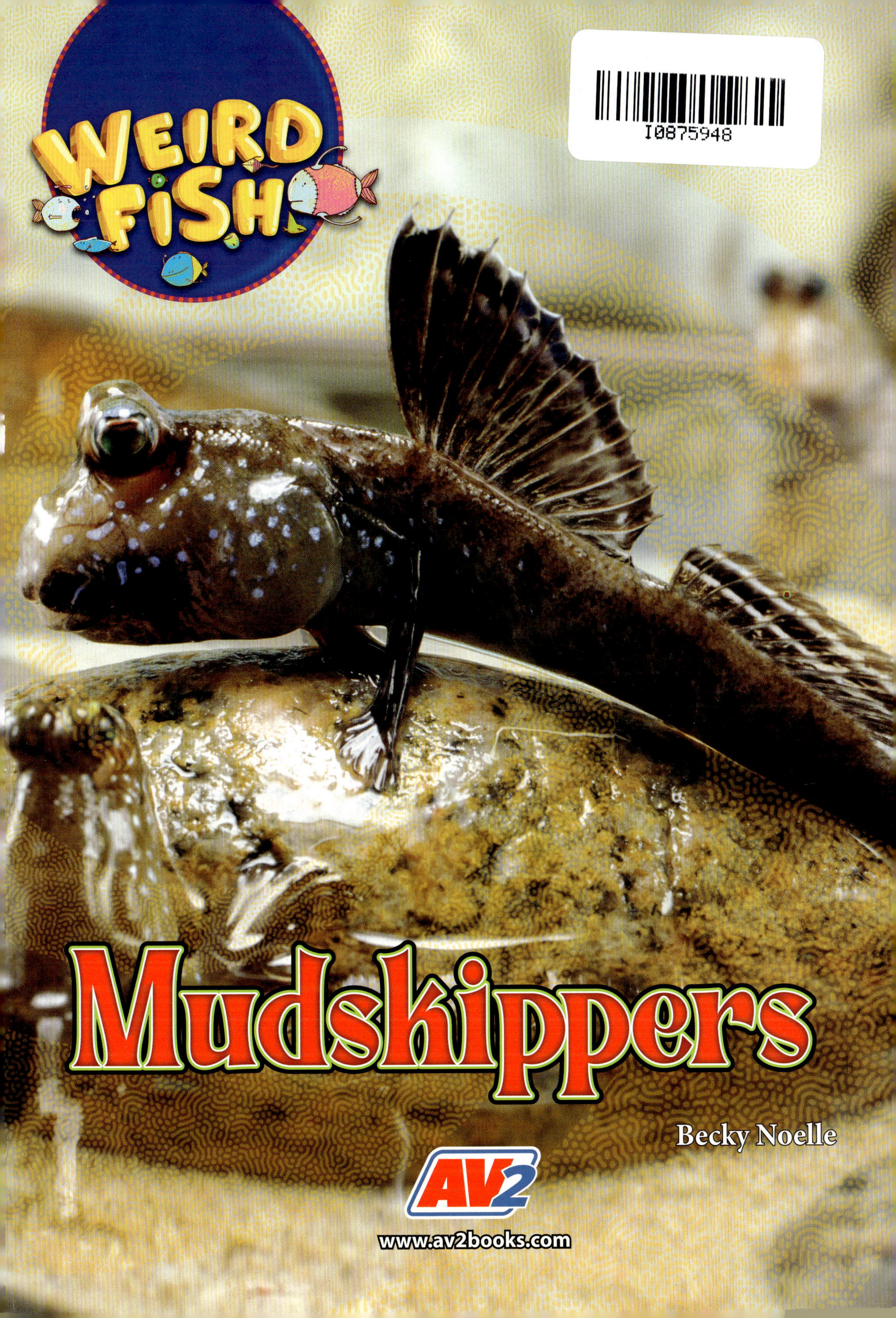

Weird Fish

Mudskippers

Becky Noelle

AV2

www.av2books.com

ISBN 978-1-7911-3320-7

WEIRD FISH

Mudskippers

CONTENTS

The Mudskipper

Have you ever seen a fish crawling on land? It might have been a mudskipper. Mudskippers live in burrows in wet, muddy **habitats**. They spend most of their time out of the water. Mudskippers use their **pectoral fins** and **pelvic fins** to walk. Some can even climb trees. Mudskippers are part of a group of fish called gobies. There are more than 30 mudskipper **species** around the world.

Mudskippers are usually brown or gray. Their fins can be brightly colored.

Atlantic Mudskipper

Scientific Name *Periophthalmus barbarus*

Diet Omnivore

Length 6 inches (15 centimeters)

Habitat Mudflats, wetlands, **estuaries**

Mudskipper Features

A mudskipper's body has many different **adaptations**. Some of these help the fish find food or stay safe. Others help it to survive in its habitat.

Skin
Mudskippers can breathe through their skin. To do this, they must keep their skin wet when they are on land by rolling in mud.

Gill Chamber
Mudskippers fill chambers connected to their **gills** with water. They close their gills to keep it in. Mudskippers can breathe on land by using this water.

Dorsal Fins

Some male mudskippers change the color of their **dorsal fins** to attract a **mate**. They also wave these fins at other males to chase them away.

Eyes

A mudskipper's eyes stick out of the top of its head. Each eye can move on its own to help the fish see. The mudskipper is the only fish that can blink.

Mudskipper Life Cycle

A male mudskipper will jump and wiggle to attract a female. The female will lay hundreds of eggs in the male's burrow. The male protects the eggs by bringing mouthfuls of fresh air to them, so they have enough oxygen to stay alive. After about a week, the male lets the burrow fill with water, so that the eggs can hatch and the baby mudskippers can swim away.

An Atlantic mudskipper can live up to 8 years.

Goby Length Comparison

Human Hand
7.2 inches (18 cm)

Dwarf Indian Mudskipper
2 inches (5 cm)

Yellow Prawn Goby
4 inches (10 cm)

Atlantic Mudskipper
6 inches (15 cm)

Mudskipper Habitats

Mudskippers live in warm coastal areas. They move between the ocean and places such as **mangrove** swamps. Mudskippers usually live in **brackish water**.

Mudskippers leave their burrows when the tide is low to find food. They return to their homes when the water rises.

Mudskipper Facts

Some blue-spotted mudskippers dig burrows that are more than 4.2 feet (1.3 meters) deep.

On the Move

Mudskippers can swim like other fish, but they also walk, climb, and even skip on land. Mudskippers rotate their pectoral fins to pull themselves along the ground. This type of movement is called crutching. Some mudskippers use their pelvic fins to climb trees. Others use them like suction cups to grab onto rocks.

Mudskippers are named because they can use their fins and tail to jump above the mud.

Mudskipper Facts

Mudskippers use their tails to jump as high as 2 feet (0.61 m) in the air.

What They Eat

Mudskippers eat small crabs, insects, and snails. They use water to suck their **prey** into their mouths. Mudskippers spit water from their mouths to cover their prey, then suck it back in. Mudskippers also eat the **algae** that grows in their muddy habitats.

Mudskippers must use water to help them swallow because they do not have tongues.

Mudskippers use their eyesight to find food. Unlike most fish, they see better on land than under the water.

Mudskipper Facts

Mudskippers can spend up to two days at a time out of the water. Their ability to eat on land helps them do this.

Staying Safe

Mudskippers have many ways to avoid being eaten. They can crawl onto land to avoid aquatic **predators**. They can jump into the water to avoid predators on land. Mudskippers also climb trees to escape danger. Their strong eyesight helps them see threats coming and flee in time.

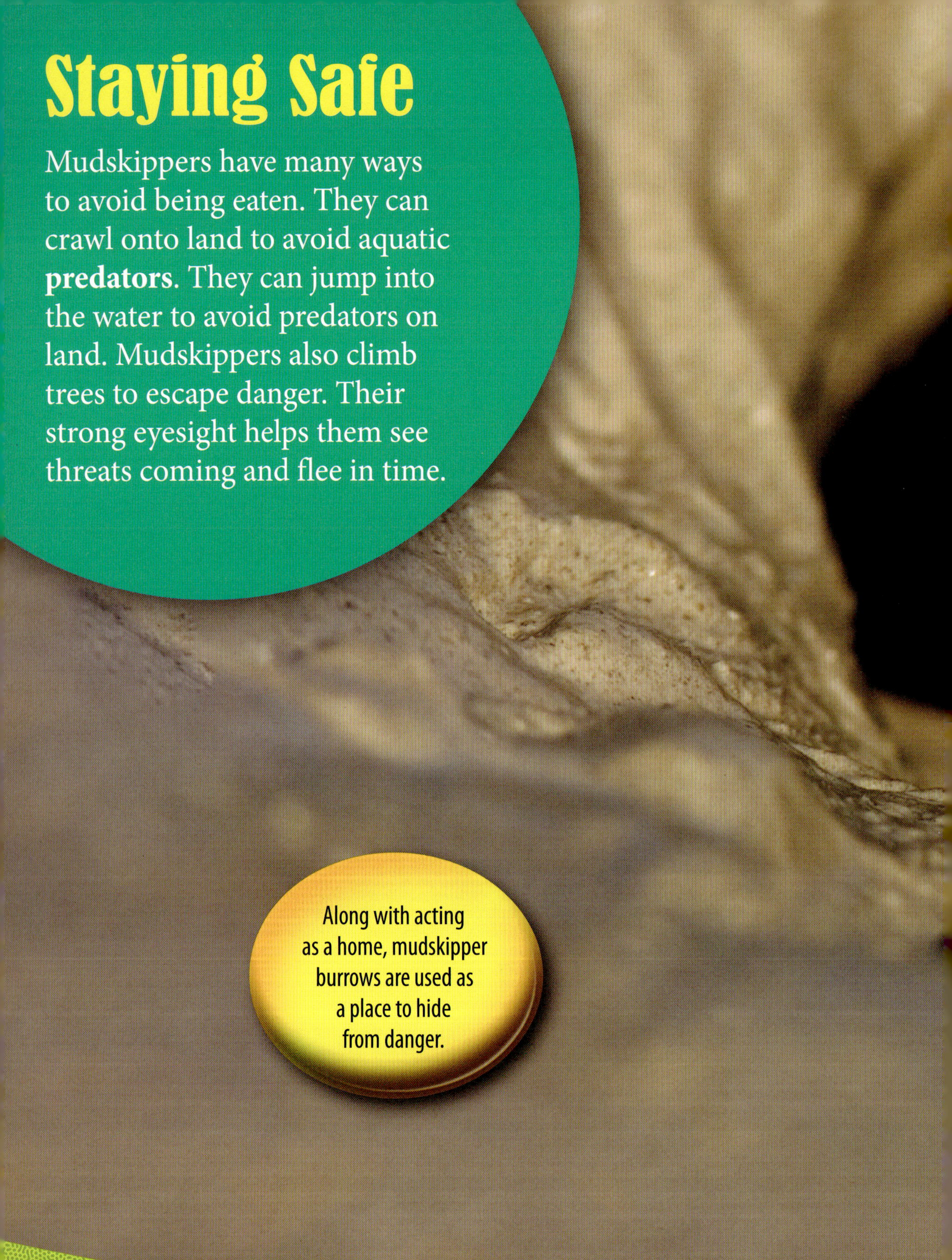

Along with acting as a home, mudskipper burrows are used as a place to hide from danger.

Viewing Mudskippers

Mudskippers can be viewed in tropical natural areas or in **aquariums**. They can be hard to see in nature because they can **camouflage** in the mud. Visitors need to go at low tide to see these fish.

World Map

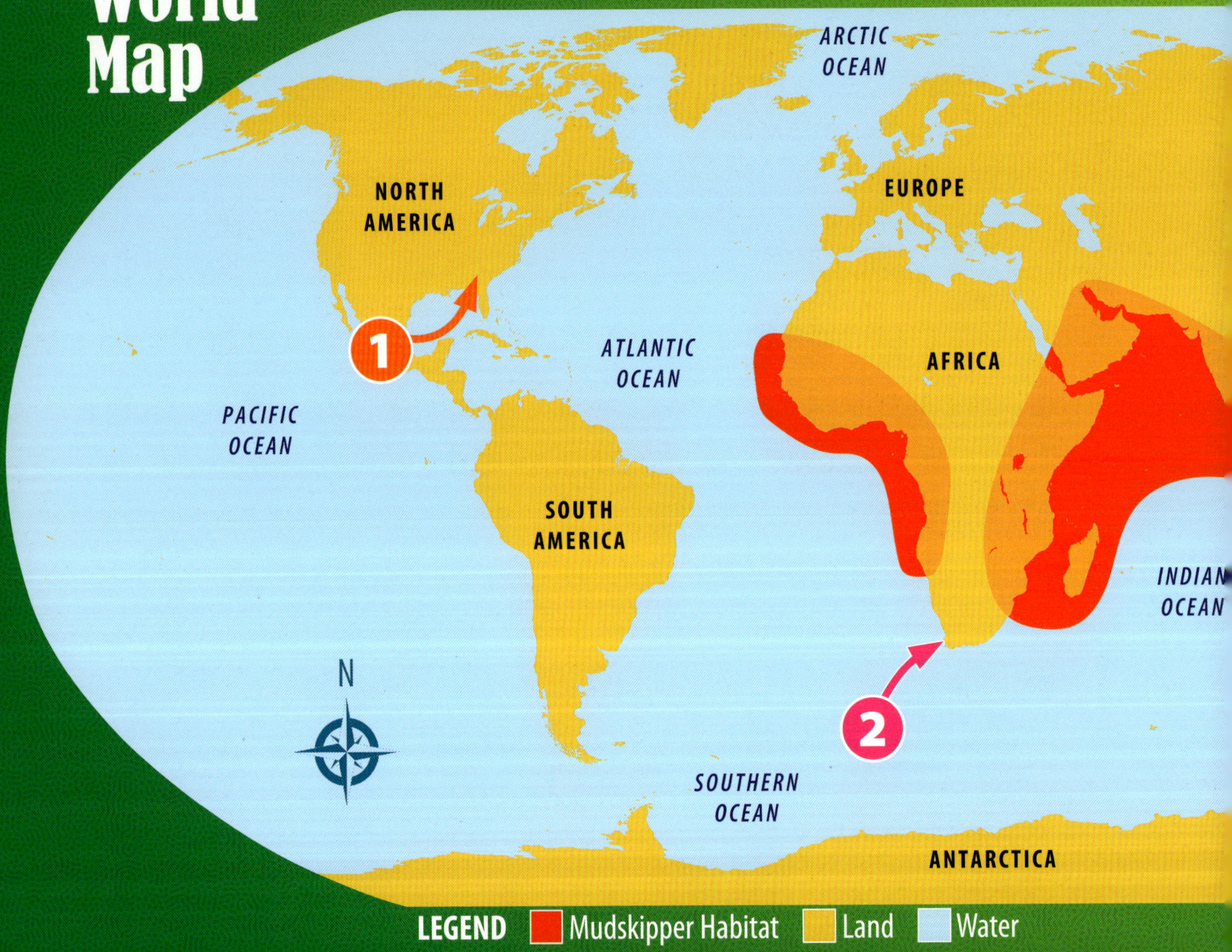

1 Atlanta, Georgia

Scientists at the Georgia Aquarium do research and teach people about ocean conservation. Aquarium visitors can view and learn about dwarf Indian mudskippers.

2 Cape Town, South Africa

The Two Oceans Aquarium in Cape Town is named after the nearby Indian and Atlantic Oceans. Barred mudskippers at the aquarium are kept separate from each other. Otherwise, these territorial fish would fight.

3 Sungei Buloh Wetland Reserve, Singapore

Mudskippers spend most of their time on muddy shores. In tropical places, visitors can spot them by walking along the beach at low tide. Sungei Buloh Wetland Reserve is one place in Singapore where mudskippers can be seen.

Protecting Mudskippers

Although mudskippers are not **endangered**, they are sensitive. They can be hurt by pollution and mangrove **deforestation**. Scientists can use mudskipper populations to test the health of a habitat. If many mudskippers live in an area, then the habitat is healthy. If there are no mudskippers, it may mean that there is too much pollution or too few mangroves.

Sensitive animals that show the health of an area are known as indicator species.

Atlantic Mudskipper Conservation Status

LC	NT	VU	EN	CR	EW	EX
Least Concern	Near Threatened	Vulnerable	Endangered	Critically Endangered	Extinct in the Wild	Extinct

Test Your Knowledge

1. What do mudskippers use to help swallow food?

2. What parts of their body do mudskippers use to breathe on land?

3. Where do mudskippers lay their eggs?

4. How high can a mudskipper jump?

5. What do mudskippers eat?

6. What do mudskippers use their pectoral fins for?

7. Where is Sungei Buloh Wetland Reserve?

8. How long is the Atlantic mudskipper?

ANSWERS
1. Water 2. Gill chambers and skin 3. In burrows
4. 2 feet (0.61 m) 5. Crabs, insects, snails, and algae 6. Walking
7. Singapore 8. 6 inches (15 cm)

Key Words

adaptations: changes in living things that make them better able to survive in their homes

algae: a group of plantlike living things that includes seaweed

aquariums: places that keep and display animals that live in water

brackish water: slightly salty water, usually found where rivers meet the ocean

camouflage: to match one's surroundings

deforestation: when many trees in a forest are cut down

dorsal fins: fins on an animal's back

endangered: close to disappearing from Earth

estuaries: the points where rivers meet oceans

gills: organs used by fish to take in oxygen from water

habitats: the places where plants or animals live

mangrove: a type of tropical forest that floods during high tide

mate: a member of a pair of animals that can reproduce or have young

pectoral fins: fins on an animal's chest

pelvic fins: fins on an animal's lower stomach area

predators: animals that hunt other animals

prey: animals that are hunted by other animals

species: groups of closely related animals with the same traits

Index

Published by AV2
276 5th Avenue
Suite 704 #917
New York, NY 10001
Website: www.av2books.com

Library of Congress Cataloging-in-Publication Data

Names: Noelle, Becky, author.
Title: Mudskippers / Becky Noelle.
Description: New York, NY : AV2, [2022] | Series: Weird fish | Includes index. | Audience: Grades 2-3
Identifiers: LCCN 2021022191 (print) | LCCN 2021022192 (ebook) | ISBN 9781791142391 (library binding) | ISBN 9781791142407 (paperback) | ISBN 9781791142414 (ebook other)
Subjects: LCSH: Mudskippers--Juvenile literature.
Classification: LCC QL638.G7 N64 2022 (print) | LCC QL638.G7 (ebook) | DDC 597.7--dc23
LC record available at https://lccn.loc.gov/2021022191
LC ebook record available at https://lccn.loc.gov/2021022192

Printed in Guangzhou, China
1 2 3 4 5 6 7 8 9 0 25 24 23 22 21

062021
101120

Art Director: Terry Paulhus Project Coordinator: John Willis

Every reasonable effort has been made to trace ownership and to obtain permission to reprint copyright material. The publisher would be pleased to have any errors or omissions brought to its attention so that they may be corrected in subsequent printings.

The publisher acknowledges Alamy, Getty Images, Minden Pictures, and Shutterstock as the primary image suppliers for this title.